SCHOLAST

Write & Draw Math

Kindergarten

by Mary Rosenberg

New York • Toronto • London
Auckland • Sydney • New Delhi • Hong Kong

Editor: Maria L. Chang
Cover design: Tannaz Fassihi
Interior design: Grafica, Inc.
Photos © Shutterstock: cover (FotoDuets), (Hazal Ak), (MuratGungut); 10 ,16 (Martial Red);
illustrations 55, 58, 60 Mattia Cerato.

Scholastic Inc., 557 Broadway, New York, NY 10012
ISBN: 978-1-338-31436-6
Copyright © 2020 by Mary Rosenberg
Published by Scholastic Inc.
All rights reserved. Printed in the U.S.A.
First printing, January 2020.

1 2 3 4 5 6 7 8 9 10   40   25 24 23 22 21 20

Table of Contents

Introduction

Welcome to *Write & Draw Math!* This book offers engaging, ready-to-use activities that help children develop flexible thinking skills about math. Flexible thinking encourages children to "think outside the box," enabling them to apply learning in new ways, brainstorm how to solve word problems, shift to a new strategy when the current one is not working, and understand abstract concepts. They learn to look at the same math problem through different lenses and explore various ways to solve it—employing a concrete method (using manipulatives), a representational method (drawing objects), or an abstract method (writing a math equation). In fact, several activities ask children to tackle a problem using at least two of these methods.

The activities in *Write & Draw Math* are ideal for presenting to the whole class or to a small group. When introducing a new activity, model for children how to do it and then monitor them while they work. Because kindergartners are just learning the important foundational skills of math, it is essential that they have an adult (teacher, instructional assistant, or parent volunteer) to guide them through the activities and to clarify any questions or mathematical misunderstandings they may have.

What makes the problems in this book unique is that there is no one correct answer. This allows children to make and share a variety of math problems and solutions using the same activity page. Since each activity uses unique numbers each time, children can complete the same page multiple times. As they become more adept at manipulating small numbers, challenge them to repeat the same activity using larger numbers. Encourage children to describe their answers using mathematical language—an important part of today's rigorous standards that is often overlooked in math instruction.

Many of the activities require the use of common classroom materials, such as:

- number cubes or dice (6-sided, as well as dice with up to 20 sides)
- small stamps and stamp pads or small stickers
- pattern blocks

- drinking straws cut up in different lengths
- different kinds of counters (teddy-bear counters, linking cubes, pennies, beans, buttons)
- place-value blocks
- paper clips for spinners

Some activities require the use of a spinner, which is provided on the page. To use the spinner, have a child place a paper clip on the spinner and use a sharpened pencil to hold one end of the paper clip in place (see right). The child then flicks a finger to make the paper clip spin.

So, what are you waiting for? Get your kindergartners started on a fun, learning-filled journey that will build their mathematical reasoning and critical thinking skills!

Math Standards Correlations

The activities in this book meet the following core standards in mathematics.

COUNTING & CARDINALITY

A. Know number names and the count sequence.

CC.A.1 Count to 100 by ones and by tens.

CC.A.2 Count forward beginning from a given number within the known sequence.

CC.A.3 Write numbers from 0 to 20. Represent a number of objects with a written numeral 0–20.

B. Count to tell the number of objects.

CC.B.4 Understand the relationship between numbers and quantities; connect counting to cardinality.

CC.B.5 Count to answer "how many?" questions about as many as 20 things arranged in a line, a rectangular array, or a circle, or as many as 10 things in a scattered configuration; given a number from 1–20, count out that many objects.

C. Compare numbers.

CC.C.6 Identify whether the number of objects in one group is greater than, less than, or equal to the number of objects in another group, e.g., by using matching and counting strategies.

CC.C.7 Compare two numbers between 1 and 10 presented as written numerals.

OPERATIONS & ALGEBRAIC THINKING

A. Understand addition as putting together and adding to, and understand subtraction as taking apart and taking from.

OA.A.1 Represent addition and subtraction with objects, fingers, mental images, drawings, sounds, acting out situations, verbal explanations, expressions, or equations.

OA.A.2 Solve addition and subtraction word problems, and add and subtract within 10, e.g., by using objects or drawings to represent the problem.

OA.A.3 Decompose numbers less than or equal to 10 into pairs in more than one way, e.g., by using objects or drawings, and record each decomposition by a drawing or equation.

OA.A.4 For any number from 1 to 9, find the number that makes 10 when added to the given number.

OA.A.5 Fluently add and subtract within 5.

NUMBER & OPERATIONS IN BASE TEN

A. Work with numbers 11–19 to gain foundations for place value.

NBT.A.1 Compose and decompose numbers from 11 to 19 into ten ones and some further ones, e.g., by using objects or drawings, and record each composition or decomposition by a drawing or equation; understand that these numbers are composed of ten ones and one, two, three, four, five, six, seven, eight, or nine ones.

MEASUREMENT & DATA

A. Describe and compare measurable attributes.

MD.A.1 Describe measurable attributes of objects, such as length or weight. Describe several measurable attributes of a single object.

MD.A.2 Directly compare two objects with a measurable attribute in common, to see which object has "more of"/"less of" the attribute, and describe the difference.

B. Clarify objects and count the number of objects in each category.

MD.B.3 Classify objects into given categories; count the number of objects in each category and sort the categories by count.

GEOMETRY

A. Identify and describe shapes.

G.A.1 Describe objects in the environment using names of shapes, and describe the relative positions of these objects using terms such as *above*, *below*, *beside*, *in front of*, *behind*, and *next to*.

G.A.2 Correctly name shapes regardless of their orientations or overall size.

G.A.3 Identify shapes as two-dimensional or three-dimensional.

B. Analyze, compare, create, and compose shapes.

G.B.4 Analyze and compare two- and three-dimensional shapes, in different sizes and orientations, using informal language to describe their similarities, differences, parts, and other attributes.

G.B.5 Model shapes in the world by building shapes from components and drawing shapes.

G.B.6 Compose simple shapes to form larger shapes.

Write & Draw Math: Kindergarten © Mary Rosenberg, Scholastic Inc.

#1 **Name:** _____

1. Count to 100. How many different ways can you do it? Show them.

2. Describe your favorite way to count to 100.

_____

_____

★ #2 **Name:** _____

1. Count by 10s. Write the missing numbers.

1	2	3	4	5	6	7	8	9	
11	12	13	14	15	16	17	18	19	
21	22	23	24	25	26	27	28	29	
31	32	33	34	35	36	37	38	39	
41	42	43	44	45	46	47	48	49	
51	52	53	54	55	56	57	58	59	
61	62	63	64	65	66	67	68	69	
71	72	73	74	75	76	77	78	79	
81	82	83	84	85	86	87	88	89	
91	92	93	94	95	96	97	98	99	

2. Circle your favorite number from the ones you wrote. Describe the number. Use tens and ones.

_____

_____

Write & Draw Math: Kindergarten © Mary Rosenberg, Scholastic Inc.

#3 Name: _____

1. Start with a number greater than 0.
Write the next numbers in the boxes.

2. Circle one of the numbers. Draw items to show that number.

3. Tell about the items you drew.

_____

_____

#4 Name: _____

Sam had fewer than 8 teddy-bear counters.
Sam then counted out 12 more counters.
How many teddy-bear counters could Sam have?

1. Tell about Sam's counters. Write the math problem.

_____

_____

2. Draw a picture to explain your answer.

3. Tell about your answer.

_____

_____

Write & Draw Math: Kindergarten © Mary Rosenberg, Scholastic Inc.

#5 **Name:** _____

1. Pick a number from 11 to 19.
   Write the number in the box.

2. Color that number of circles.

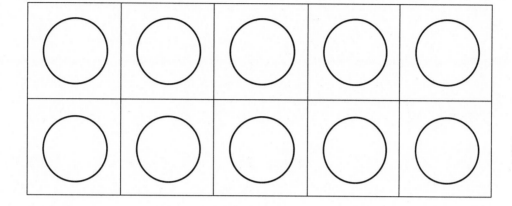

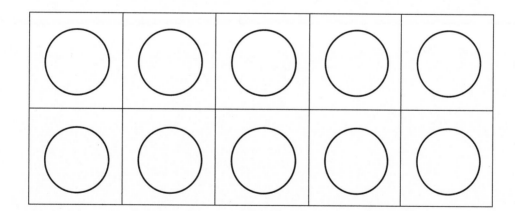

3. Describe the number you made. Use tens and ones.

_____

_____

_____

#6 **Name:** _____

1. Pick a number from 1 to 20.
Write the number in the box.

2. Draw that number of items.

3. Tell about the number you picked.

_____

_____

Write & Draw Math: Kindergarten © Mary Rosenberg, Scholastic Inc.

#7 Name: _____

1. Spin the spinner. Write the number in the box.

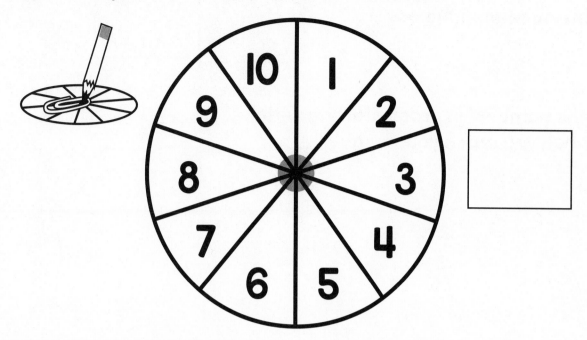

2. Draw the matching number of circles.

3. What number did you spin? _____

4. What number did you write in the box? _____

5. How many circles did you make? _____

Write & Draw Math: Kindergarten © Mary Rosenberg, Scholastic Inc.

#8 Name: _____

1. Pick a number between 1 and 20.
   Write the number in the box.

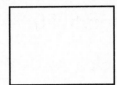

2. Write the numbers in order, starting with 1.
   Stop when you reach your number.

3. Draw a circle or square under each number.

   How many shapes did you make? _____

4. What would the next number be? _____

Write & Draw Math: Kindergarten © Mary Rosenberg, Scholastic Inc.

#9 Name: _____

1. Take a handful of teddy-bear counters.

2. Draw a picture to show the number of counters.

3. Number each counter, starting with 1.

How many counters did you take? _____

4. Repeat Steps 1 to 3.

How many counters did you take this time? _____

5. Compare the numbers of teddy-bear counters.

_____ counters are more than _____ counters.

_____ counters are less than _____ counters.

_____ counters are the same as _____ counters.

#10 Name: _____

1. Roll a number cube. Use counters to cover the matching number of bears. How many more do you need to make 10?

I rolled a _____ . I need _____ more to make 10.

_____ + _____ = 10

2. Repeat Step 1.

I rolled a _____ . I need _____ more to make 10.

_____ + _____ = 10

3. Describe one of the ways you made 10.

_____

_____

_____

_____

Write & Draw Math: Kindergarten © Mary Rosenberg, Scholastic Inc.

#11 Name: _____

1. Roll a number cube. Count out that number of counters. (You can use cubes, blocks, beans, buttons, or pennies.)

2. Trace each item below. Number each item. Tell how many in all.

3. Tell about your answer.

_____

_____

_____

#12 Name: _____

1. Spin the spinner.

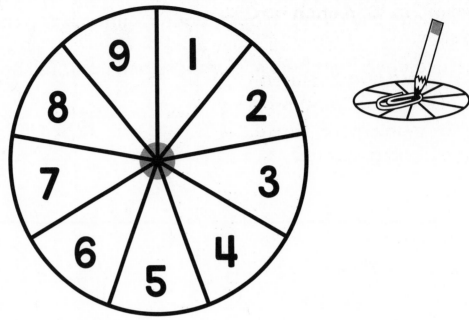

2. Draw the matching number of circles.

3. What number is one more? _____

4. How do you know?

_____

_____

Write & Draw Math: Kindergarten © Mary Rosenberg, Scholastic Inc.

#13 Name: _____

1. Write a different number from 1 to 10 in each box below.

2. For each number, spin the spinner. Count on that many more. Circle the ending number.

For example:

Rosie picked the number 8.
She spun the number 4.

8	9 10 11 ⑫

	_____

	_____

	_____

3. What was the largest sum? _____

4. What was the smallest sum? _____

#14 **Name:** _____

1. Spin the spinner. Write the number in the box.

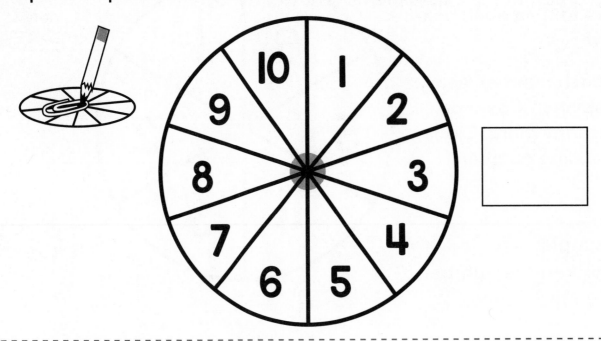

2. Show the number in different ways.

3. Describe one of the ways you showed the number.

_____

_____

Write & Draw Math: Kindergarten © Mary Rosenberg, Scholastic Inc.

#15 Name: _____

1. Drop a handful of cubes in this box. Don't move them.

2. Count how many cubes are in the box.

There are _____ cubes in the box.

3. Now, arrange the cubes into one line. Count them.

There are _____ cubes in one line.

4. Finally, put the cubes into two rows. How many cubes are there?

There are _____ cubes in two rows.

5. Did the numbers change? Why?

_____

_____

#16 Name: _____

1. Roll a number cube.
   Write the number in the box.

2. Draw the matching number of circles.

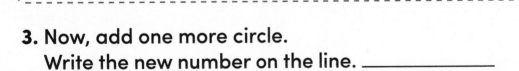

3. Now, add one more circle.
   Write the new number on the line. _____

4. Write a story problem about the numbers.

_____

_____

_____

Write & Draw Math: Kindergarten © Mary Rosenberg, Scholastic Inc.

Write & Draw Math: Kindergarten © Mary Rosenberg, Scholastic Inc.

#17   Name: _____

1. Grab a handful of red counters. Place one counter on each line. Write the total number in the box.

____ ____ ____ ____ ____ ____ ____ ____

2. Repeat Step 1 with blue counters.

____ ____ ____ ____ ____ ____ ____ ____

3. Compare the number of red counters and blue counters. Complete each sentence.

There are _____ red counters.
(number)

There are _____ blue counters.
(number)

There are **more** _____ counters than _____ counters.
(color) (color)

There are **fewer** _____ counters than _____ counters.
(color) (color)

There are _____ **more** _____ counters
(number) (color)
than _____ counters.
(color)

There are _____ **fewer** _____ counters
(number) (color)
than _____ counters.
(color)

23

#18 Name: _____

1. How many counters does Jack have? Roll a number cube to see. Put that many counters in Jack's circle.

2. How many counters does Jill have? Roll the number cube to see. Put that many counters in Jill's circle.

Jack's counters

Jill's counters

Jack has _____ counters.

Jill has _____ counters.

3. Who has the greater number of counters? _____

4. How do you know?

_____

_____

Write & Draw Math: Kindergarten © Mary Rosenberg, Scholastic Inc.

#19 Name: _____

1. Spin the spinner. Write the number in the box.
Draw the matching number of circles.

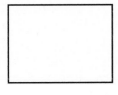

2. Repeat Step 1.

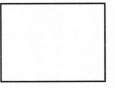

3. Write a sentence comparing the two numbers.

_____

_____

#20 **Name:** _____

1. Roll a number cube. Write the number in the first box.

2. Roll the number cube again. Write the number in the second box.

3. Draw the matching number of circles under each box.

4. Circle the larger number.

5. Compare the two numbers. Use the symbols <, >, or =.

6. Repeat Steps 1 and 2.

Write & Draw Math: Kindergarten © Mary Rosenberg, Scholastic Inc.

#21 **Name:** _____

Mario wants to break down the number 8 into two smaller numbers.

Show different ways he could break down, or decompose, 8.

1. Draw a picture.

2. Write the addition problem for each picture.

_____ + _____ = 8	_____ + _____ = 8
_____ + _____ = 8	_____ + _____ = 8

Write & Draw Math: Kindergarten © Mary Rosenberg, Scholastic Inc.

#22 **Name:** _____

Steph wants to break down the number 8 into
two smaller numbers.

Show different ways she could break down, or decompose, 8.

1. Draw pictures.

2. Write the subtraction problem for each picture.

8 – _____ = _____   8 – _____ = _____

8 – _____ = _____   8 – _____ = _____

Write & Draw Math: Kindergarten © Mary Rosenberg, Scholastic Inc.

#23 **Name:** _____

1. Write a number sentence with the numeral 5 in it.

2. Describe your number sentence.

_____

_____

_____

#24 Name: _____

There is more than 1 bear in a cave. There are fewer than 10 bears in the cave. How many bears could be in the cave?

1. Draw a picture of the problem.

There are _____ bears in the cave.

2. Explain your answer.

_____

_____

Write & Draw Math: Kindergarten © Mary Rosenberg, Scholastic Inc.

#25 **Name:** _____

Mya has 10 blocks in all. Some blocks are blue.
The rest are green. How many of the blocks could be blue?
How many could be green?

1. Draw a picture of the problem.

There could be _____ blue blocks and _____ green blocks.

2. Explain how you solved this problem.

_____

_____

#26 Name: _____

Tara started counting from 25.

1. Write three numbers Tara would say to get to the number 50.

2. Explain how you solved this problem.

_____

_____

Write & Draw Math: Kindergarten © Mary Rosenberg, Scholastic Inc.

#27 **Name:** _____

1. Write three addition problems with a sum of 10.

_____ + _____ = 10

_____ + _____ = 10

_____ + _____ = 10

2. Explain how you solved this problem.

_____

_____

_____

#28 **Name:** _____

1. Write three subtraction problems with a difference of 5.

_____ – _____ = 5

_____ – _____ = 5

_____ – _____ = 5

2. Explain how you solved this problem.

_____

_____

_____

Write & Draw Math: Kindergarten © Mary Rosenberg, Scholastic Inc.

#29 **Name:** _____

The sum of two numbers is 10.

1. What could the two numbers be?

2. Explain how you solved this problem.

_____

_____

_____

#30 Name: _____

The difference between two numbers is 5.

1. What could the two numbers be?

2. Explain how you solved this problem.

_____

_____

_____

Write & Draw Math: Kindergarten © Mary Rosenberg, Scholastic Inc.

#31 **Name:** _____

1. Spin the spinner.

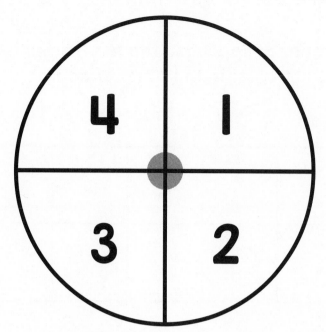

2. Color in that number of spaces in the five frame. Use a red crayon. Use a blue crayon to color in the other spaces.

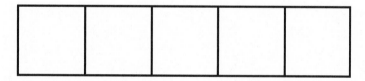

3. Write the addition problem.

_____ + _____ = _____

4. Tell about the addition sentence you made.

_____

_____

_____

#32 Name: _____

1. Roll a number cube.

2. Color in that number of spaces in the ten frame. Use a red crayon. Use a blue crayon to color in the other spaces.

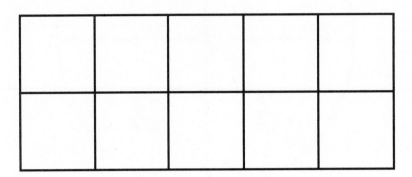

3. Write the addition problem.

_____ + _____ = _____

4. Tell about the addition sentence you made.

_____

_____

_____

_____

Write & Draw Math: Kindergarten © Mary Rosenberg, Scholastic Inc.

#33 **Name:** _____

Jamal wants to break apart 10 into two smaller numbers.

1. Show 4 different ways Jamal could do this.

2. Tell about one of the ways.

_____

_____

_____

Write & Draw Math: Kindergarten © Mary Rosenberg, Scholastic Inc.

#34 Name: _____

1. Grab a handful of cubes.
Sort the cubes into groups of tens and ones.

2. How many tens did you make? _____

3. How many ones were left? _____

4. What number did you make? _____

#35 Name: _____

Note to Teacher: Provide a 12-, 15-, or 20-sided die.

1. Roll a die.

2. Color the ten frame to show the number you rolled.

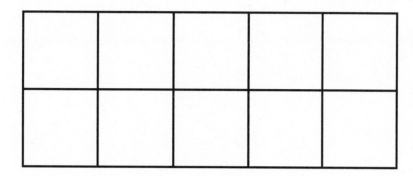

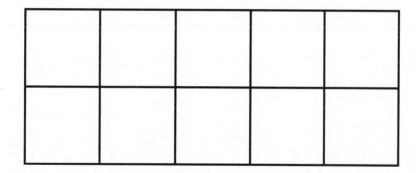

3. Describe the number you rolled. Use tens and ones.

_____

_____

_____

Write & Draw Math: Kindergarten © Mary Rosenberg, Scholastic Inc.

#36 Name: _____

Andy counted a total of 10 pigs and cows.

1. **Show the number of pigs and cows Andy saw.**
   Use linking cubes. Write the math problem.

2. **Tell how you figured out how many pigs and cows Andy saw.**

_____

_____

_____

Write & Draw Math: Kindergarten © Mary Rosenberg, Scholastic Inc.

⭐ #37 Name: _____

Note to Teacher: Provide a 12-, 15-, or 20-sided die.

1. Roll a die. Write the number in the box.

2. Show the number using place-value blocks.
   Draw a quick picture showing the place-value blocks you used.

3. How many tens did you make? _____

4. How many ones did you make? _____

Write & Draw Math: Kindergarten © Mary Rosenberg, Scholastic Inc.

#38   **Name:** _____

Note to Teacher: Set a timer to 30 seconds.

1. Put a stamp or sticker in each square.
   How many squares can you fill in 30 seconds?

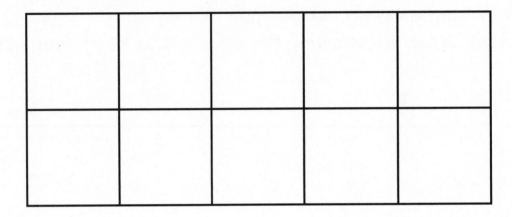

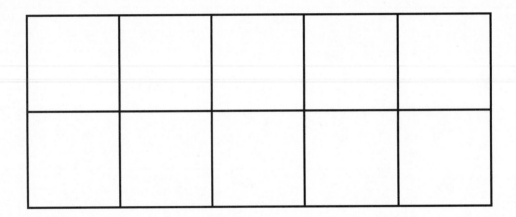

2. How many tens did you stamp or sticker? _____

3. How many ones did you stamp or sticker? _____

4. Tell about the number you made.

_____

_____

Write & Draw Math: Kindergarten © Mary Rosenberg, Scholastic Inc.

#39 Name: _____

Jan is thinking of a number between 0 and 20.
It has 1 as a digit.

1. Write all the possible numbers that could be Jan's number.

2. Tell about the numbers you wrote.
   Explain how they could be Jan's number.

_____

_____

_____

#40 Name: _____

1. Pick a number from 11 to 19.
   Write the number in the box.

2. Show the number on the ten frames.

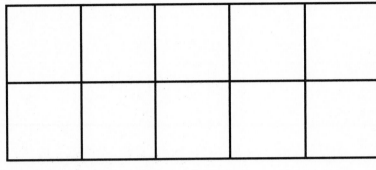

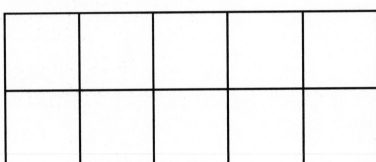

3. Draw the number of tens and ones.

Write & Draw Math: Kindergarten © Mary Rosenberg, Scholastic Inc.

#**41** **Name:** _____

Note to Teacher: Set a timer to 20 seconds.

1. Draw as many happy faces as you can in 20 seconds.

2. Write the number of happy faces you made.

3. Show the number on the ten frames.

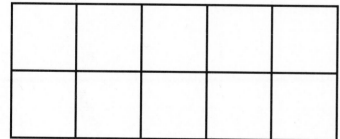 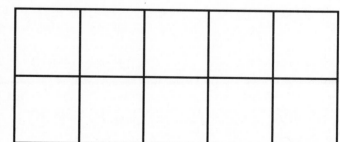

4. Write the number of tens and ones.

_____ tens _____ ones

#42 Name: _____

Note to Teacher: Provide a basket of straws cut to different lengths and a basket of same-size cubes (blocks, linking cubes, and so on).

1. Pick two straws and match up the ends.
   Draw the length of each straw.

Straw 1

Straw 2

2. Use words to compare the straws.

Straw 1 is _____ than Straw 2.
                 (longer/shorter)

Straw 2 is _____ than Straw 1.
                 (longer/shorter)

3. Use cubes to measure the length of each straw.

Straw 1 is _____ cubes long.
              (number)

Straw 2 is _____ cubes long.
              (number)

Straw 1 is _____ cubes _____ than Straw 2.
              (number)   (longer/shorter)

Straw 2 is _____ cubes _____ than Straw 1.
              (number)   (longer/shorter)

Write & Draw Math: Kindergarten © Mary Rosenberg, Scholastic Inc.

#43 Name: _____

1. Draw a picture of you standing next to your friend.
   Who is taller? Who is shorter?

2. Use words to compare your height to your friend's height.

I am _____ than my friend.
         (taller/shorter)

My friend is _____ than I am.
             (taller/shorter)

#44 Name: _____

1. Build two towers of different heights.
   Use blocks of two different colors.

2. Compare the heights of the two towers.

3. Tell about the towers you made.

_____

_____

_____

Write & Draw Math: Kindergarten © Mary Rosenberg, Scholastic Inc.

1. Pick 10 buttons from a box.

2. Sort the buttons. You can sort by size, color, shape, or design.

Group 1	**Group 2**
Group 3	**Group 4**

3. Describe how you sorted the buttons.

_____

_____

_____

#46 Name: _____

1. Find different shapes in the classroom.

2. Count the number of sides and corners on each shape.

Shapes	Classroom Item	Number of Sides	Number of Corners
Triangle			
Square			
Rectangle			
Circle			
Trapezoid			

Write & Draw Math: Kindergarten © Mary Rosenberg, Scholastic Inc.

#**47** Name: _____

1. Find different shapes in the classroom.

2. Count the number of sides and corners on each shape.

Shapes	Classroom Item	Number of Sides	Number of Corners
Sphere			
Cube			
Rectangular prism			
Pyramid			
Cylinder			

#48 Name: _____

1. Use <u>three</u> pattern blocks to create a new shape. Trace the shapes you used. How many sides and corners does your new shape have?

My new shape has _____ sides and _____ corners.

2. Use <u>four</u> pattern blocks to create a new shape. Trace the shapes you used. How many sides and corners does your new shape have?

My new shape has _____ sides and _____ corners.

3. Use <u>five</u> pattern blocks to create a new shape. Trace the shapes you used. How many sides and corners does your new shape have?

My new shape has _____ sides and _____ corners.

#49   Name: _____

1. How many different shapes can you find in the kid's bedroom?

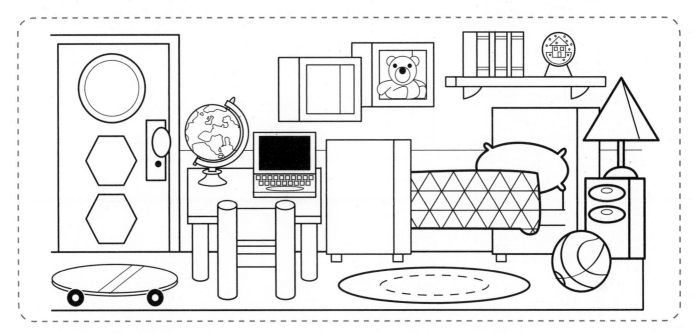

2. List the items in the picture. Write the shape of each item.

Bedroom Item	Shape
Example: door knob	oval

1. Pick a 2-dimensional shape.

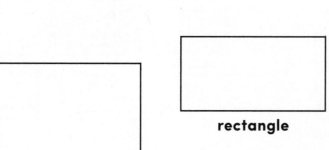

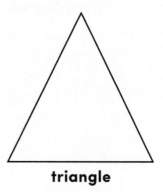

triangle

square

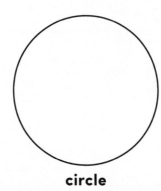

circle

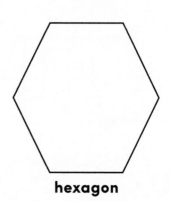

hexagon

2. Write a riddle about the shape.
Ask a friend to find the shape that answers the riddle.

My riddle:

_____

_____

_____

Friend's answer:

_____

Write & Draw Math: Kindergarten © Mary Rosenberg, Scholastic Inc.

Name: _____

1. Pick a 3-dimensional shape.

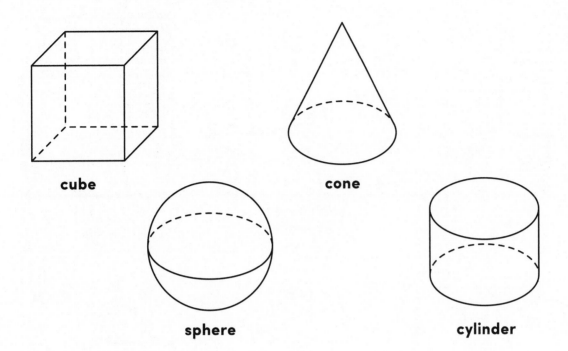

2. Write a riddle about the shape.
Ask a friend to find the shape that answers the riddle.

My riddle:

_____

_____

_____

Friend's answer:

_____

#52 Name: _____

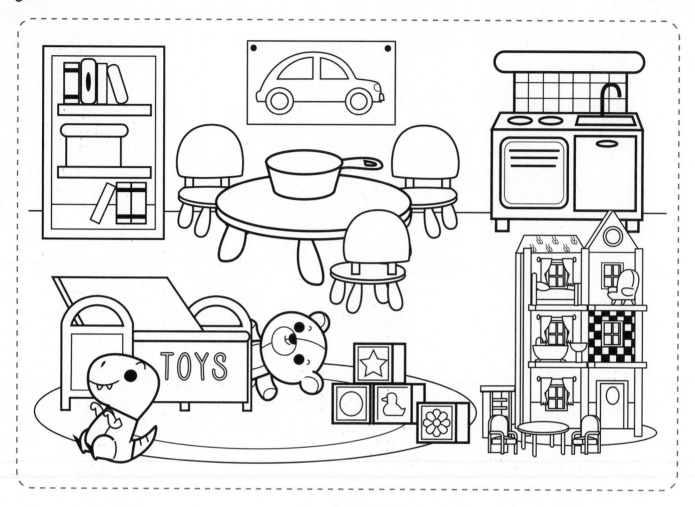

Describe the location of three different items in the playroom.
Use some of the words below.

Word Bank

above	behind	beside	below	in front of	next to

1. _____

2. _____

3. _____

Write & Draw Math: Kindergarten © Mary Rosenberg, Scholastic Inc.

#53 Name: _____

---

1. How many different triangles can you make?

What makes a shape a triangle?

_____

_____

---

2. How many different rectangles can you make?

What makes a shape a rectangle?

_____

_____

#**54** **Name:** _____

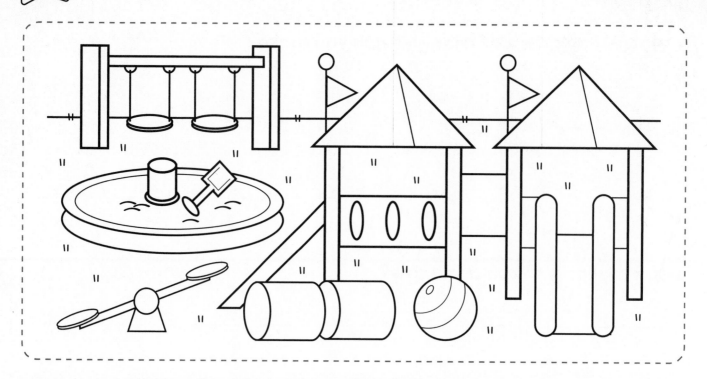

1. Do you see 2-dimensional (flat) shapes and 3-dimensional (solid) shapes in the picture? Make a list of them below.

2-Dimensional Shapes	3-Dimensional Shapes

2. Describe your favorite shape.
    Is it a 2-dimensional or 3-dimensional shape?

_____

_____

Write & Draw Math: Kindergarten © Mary Rosenberg, Scholastic Inc.

#55 **Name:** _____

1. Look at these two different shapes.

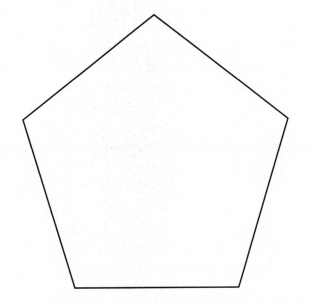 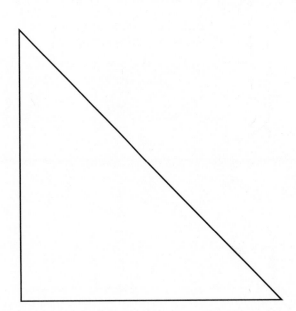

2. How are they alike?

They are alike because they both _____

_____ .

3. How are they different?

They are different because one shape _____

_____

and the other shape _____

_____ .

Name: _____

1. Put two or more of the same pattern blocks together.

2. Trace the blocks. What new shapes did you make?

I put _____ _____ together and made
　　　(number)　　　　(shapes used)

a _____ .

I put _____ _____ together and made
　　　(number)　　　　(shapes used)

a _____ .

Write & Draw Math: Kindergarten © Mary Rosenberg, Scholastic Inc.

#57 Name: _____

1. Use pattern blocks to compose a triangle, square, rectangle, or hexagon.

2. Trace the blocks you used.

3. Tell about the shape you made. What blocks did you use?

_____

_____

_____

Write & Draw Math: Kindergarten © Mary Rosenberg, Scholastic Inc.

#58 Name: _____

1. Use pattern blocks to make a hexagon in as many ways as possible.

2. Trace the pattern blocks you used to make each hexagon.

3. Which pattern blocks did you use to make your favorite hexagon?

diamond   rhombus   square

trapezoid   triangle

Write & Draw Math: Kindergarten © Mary Rosenberg, Scholastic Inc.